Look up at the sky on a sunny day. Think of yourself standing on planet Earth spinning in space. Eight more planets are also spinning in space. And all nine planets are circling around a star called the Sun.

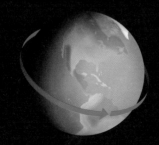

Earth

Mars

Each planet follows its own path around the Sun. The path that each planet travels is called its *orbit*. The time it takes to complete an orbit is one year on that planet.

Each planet also *rotates*, or spins like a top, as it moves through its orbit. The time it takes to complete a rotation is one day on that planet.

Each planet pulls on objects that are on or near the planet. The pull is called *gravity*. The more matter a planet has, the greater the pull of gravity. The greater the pull, the more something weighs on that planet.

The nine planets that orbit the Sun are: Mercury, Venus, Earth, Mars, Jupiter, Saturn, Uranus, Neptune, and Pluto. They make up our solar system.

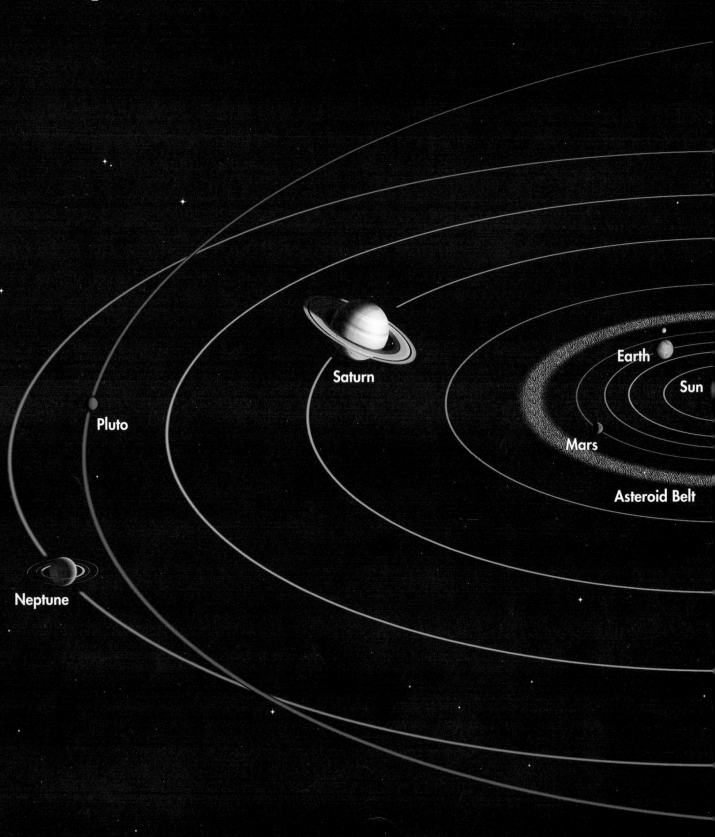

The solar system also includes moons that go around some planets, asteroids, which are big chunks of rock, and comets, which are balls of ice and dust.

Our planet **Earth** is a globe with three main layers: crust, mantle, and core. We live on the crust, which is made of hard rock. Most of the crust is covered with great bodies of water, such as oceans, lakes, and rivers. Because of all this water, people have nicknamed Earth the "blue planet."

A layer of air called the atmosphere surrounds the Earth. The atmosphere has two main gases—nitrogen and oxygen. Without air to breathe, we could not live on planet Earth.

Below the Earth's crust lie the mantle and core, which are made up of rock and metal. In some places these layers are so hot that the rock and metal have melted into liquids. If the Earth were an orange, the crust and atmosphere would only be as thick as its skin. All the rest would be mantle and core.

Would you like to tour the planets on an imaginary spaceship? Climb aboard. Buckle up. Here we go!

First stop: the **Moon**. On the Moon, everything looks very strange. The sky is always black. Mountains, valleys, and big craters cover the Moon's surface. There is no sign of life anywhere.

The rocks on the Moon are much like the rocks on Earth. But there is no atmosphere here. To walk on the Moon, astronauts need space suits. The suits have tanks of air for the astronauts to breathe.

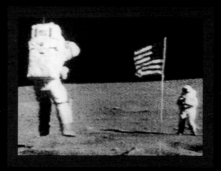

Because the Moon is smaller than Earth, the pull of gravity is less. Humans weigh less on the Moon than on Earth. The astronauts can jump higher and farther than on Earth.

The Moon does not make any light of its own. The moonlight we see from Earth is really light from the Sun hitting the Moon and bouncing off, like light bouncing off a mirror. The planets also shine because of light from the Sun.

Fire up the rockets and speed the spaceship toward the **Sun**, the star at the center of our solar system. Like other stars, the Sun is a gigantic ball of superhot gases. The Sun sends out tremendous amounts of heat and light energy. The energy is powerful enough to reach every planet in the solar system.

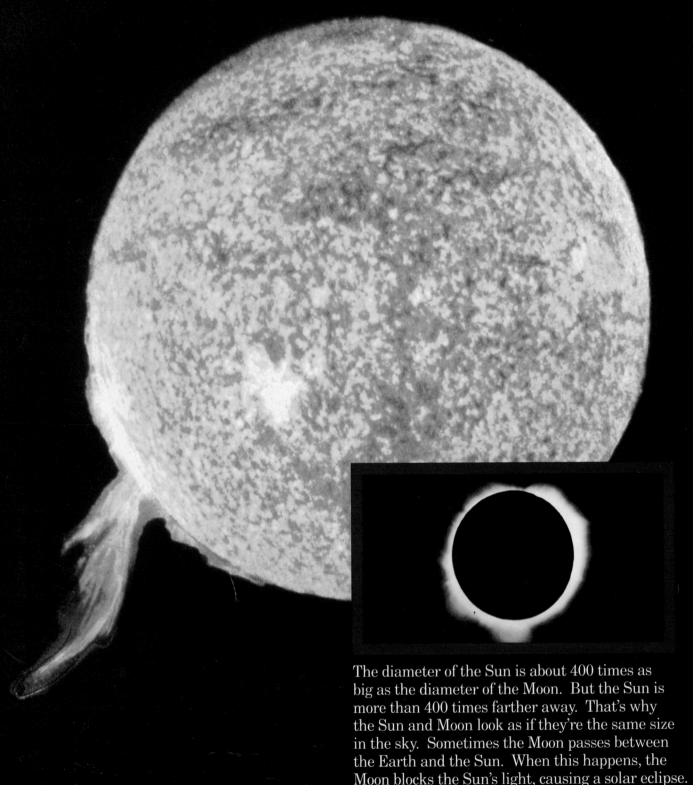

The diameter of the Sun is about 400 times as big as the diameter of the Moon. But the Sun is more than 400 times farther away. That's why the Sun and Moon look as if they're the same size in the sky. Sometimes the Moon passes between the Earth and the Sun. When this happens, the Moon blocks the Sun's light, causing a solar eclipse.

Guide the spaceship away from the Sun and toward **Mercury**, the closest planet to the Sun. Because Mercury is so near the Sun, it is unbelievably hot. During the day, the temperature climbs to 800 degrees Fahrenheit.

One rotation = 59 Earth days

One orbit = 88 Earth days

The surface of Mercury looks like the surface of our Moon. Like the Moon, Mercury has many steep cliffs and deep craters. The craters were caused by asteroids or comets that crashed down from outer space.

Mercury orbits faster than any other planet. It zips around the Sun at over 100,000 miles an hour! That gives Mercury the shortest year in the solar system. At the same time, Mercury rotates very slowly. That gives Mercury one of the longest days in the solar system.

Pilot the spaceship millions of miles toward **Venus**. Venus is Earth's nearest neighbor in space. The first thing you see around Venus is a thick layer of poisonous acid clouds.

The clouds on Venus trap the Sun's heat, creating what's known as a "greenhouse effect." The temperature on Venus's surface is about 864 degrees Fahrenheit. That makes it hotter than any other planet— and hotter than most ovens.

The surface of Venus, like that of Mercury, has many craters made by crashing asteroids. Thousands of volcanoes also dot the planet.

Did you know that you can see Venus from Earth? In fact, people used to think that Venus was a star. They called it the "morning star" when they saw it before sunrise, and the "evening star" when they saw it after sunset.

Pass through the orbit of Earth to find **Mars**, the fourth planet from the Sun. Like Earth, Mars is made of rock, but it looks red because of rusty iron in its soil. The huge gash across the planet is called Valles Marineris. It is 3,000 miles long, hundreds of miles wide, and 5 miles deep—making it much bigger than the Grand Canyon.

Mercury

Venus

Earth

Mars

Looking down from the spacecraft, we can see a huge volcano named Olympus Mons. This volcano is 15 miles high, which is three times the height of Mount Everest, the tallest mountain on Earth.

Mars is the last of the four earthlike inner planets. The inner planets are all solid bodies of rock and metal

Before reaching the next planet, we must rocket through the **asteroid belt**. Thousands of asteroids orbit the Sun in this space between the inner and outer planets—and sometimes a bright comet flashes through the asteroid belt.

People sometimes call asteroids "minor planets" because they are made of rock and metal and orbit around the Sun. The largest asteroid is named Ceres (SEER-eez). It is only about 600 miles in diameter. Most asteroids are less than one mile in diameter.

Now and then, asteroids fall out of orbit and fly toward Earth. Most burn up before they land. But a large asteroid may have crashed into Earth 65 million years ago. Some think the very thick dust it raised cut off the sunlight. The lack of sunlight killed the plants and then the dinosaurs.

Visit **Jupiter** next, but don't even think of landing.
The planet has no solid surface. It is made almost
completely of hot swirling gases, covered by a layer
of frozen clouds.

Earth

Jupiter

Surely the most thrilling sight on Jupiter is the

Fly past **Saturn** too. Like Jupiter, Saturn is a ball of gas, surrounded by swirling clouds and many moons. But Saturn's rings make it one of the most beautiful planets in the solar system.

As we approach the rings, we can see that they

Landing on the cold, blue-green gas of **Uranus** is also impossible. Like all planets, Uranus orbits around the Sun and rotates. But while the other planets are nearly upright, Uranus tilts to one side. For part of its orbit, its north pole points to the Sun!

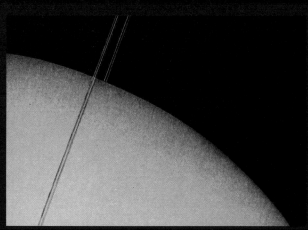

Uranus has at least 11 thin, pale gray rings. These rings are harder to see than the rings around Saturn. Some say that the rings around Uranus are made of chunks of graphite, the black material found inside pencils.

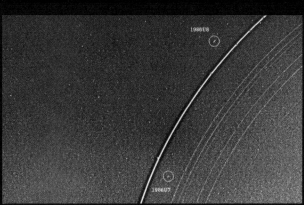

Twenty-one moons orbit Uranus. Astronomers found some of them by looking through telescopes. Another ten were discovered in 1986 by the spacecraft *Voyager II*. This photo taken by *Voyager* shows two of the newly discovered moons. Scientists have named them Ophelia and Cordelia.

We must fly more than a billion miles to get to **Neptune**, another blue-green planet. Very far from the Sun, Neptune is the last of the four giant gas planets in the solar system. Stormy winds of up to 1,200 miles an hour whip through Neptune's thick cover of clouds.

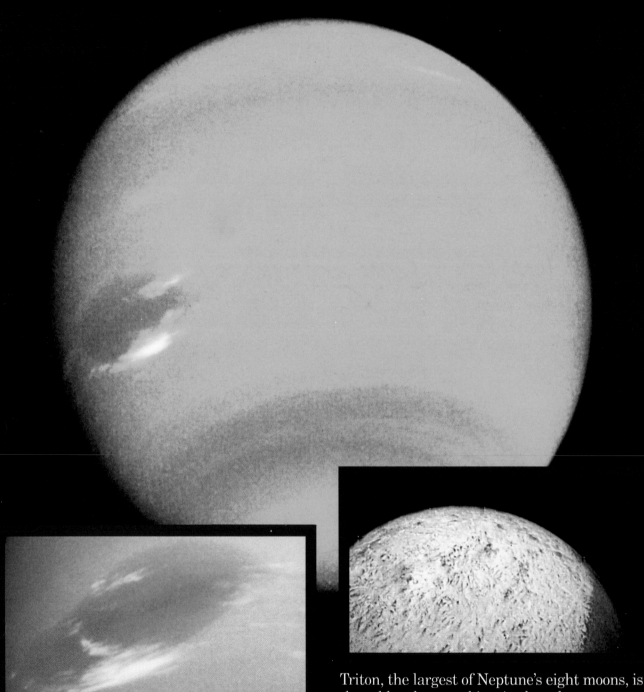

Especially violent masses of gases swirl around one huge area of Neptune. Scientists call this stormy region the Great Dark Spot. The site looks much like the Great Red Spot on Jupiter.

Triton, the largest of Neptune's eight moons, is the coldest known object in the solar system. The temperature on its surface is about 400 degrees below zero Fahrenheit. Triton may have been a comet that was captured by Neptune's gravity. Much of its surface is covered with a mixture of frozen water and ammonia.

Try to spot **Pluto**, the most distant planet, at the very edge of the solar system. Looking up at the sky from Pluto, the Sun just looks like an unusually bright star.

Pluto
(seen through
a telescope)

Charon

Sun

Pluto

Neptune

Unlike the four giant gas planets, Pluto seems to be a small solid planet like Earth. Pluto has only one moon, which is called Charon. This moon is almost half as big as the planet itself.

Pluto travels around the Sun in an oddly-shaped orbit that takes 248 Earth years to complete. For about 20 of those 248 years, Pluto is closer to the Sun than Neptune is. Then Pluto swings out and once again becomes the most distant planet from the Sun.

What is beyond Pluto? Billions and billions of stars. Each looks like a tiny spot in the night sky. But each is a giant ball of superhot gas like the Sun. And many may have planets circling around them—just like the planets in our solar system.

Our tour of the planets is now done. It's time to turn the spaceship around and head back. It was fun to travel through space. But it will be even better to return safely home to planet Earth.